Dear parents,

As a mom and as an educator, I am very excited to share the Tiger Math Workbook series with all of you. I developed this series for my two kids in elementary school, utilizing all of my knowledge and experience that I have gained while studying and working in the fields of Elementary Education and Gifted Education in South Korea as well as in the United States.

While raising my kids in the U.S., I had great disappointment and dissatisfaction about the math curriculum in the public schools. Based on my analysis, students cannot succeed in math with the current school curriculum because there is no sequential building up of fundamental skills. This is akin to building a castle on sand. So instead, I wanted to find a good workbook, but couldn't. And I also tried to find a tutor, but the price was too expensive for me. These are the reasons why I decided to make the Tiger Math series on my own.

The Tiger Math series was designed based on my three beliefs toward elementary math education.

1. It is extremely important to build foundation of math by acquiring a sense of numbers and mastering the four operation skills in terms of addition, subtraction, multiplication, and division.
2. In math, one should go through all steps in order, step by step, and cannot jump from level 1 to 3.
3. Practice math every day, even if only for 10 minutes.

If you feel that you don't know where your child should start, just choose a book in the Tiger Math series where your child thinks he/she can complete most of the material. And encourage your child to do only 2 sheets every day. When your child finishes the 2 sheets, review them together and encourage your child about his/her daily accomplishment.

I hope that the Tiger Math series can become a stepping stone for your child in gaining confidence and for making them interested in math as it has for my kids. Good luck!

Michelle Y. You, Ph.D.
Founder and CEO of Tiger Math

ACT scores show that only one out of four high school graduates are prepared to learn in college. This preparation needs to start early. In terms of basic math skills, being proficient in basic calculation means a lot. Help your child succeed by imparting basic math skills through hard work.

Sungwon S. Kim, Ph.D.
Engineering professor

Level C – 2: Plan of Study

Goal A	Practice subtracting single digit numbers from the numbers in between 15 and 20. (Week 1 ~ 3)
Goal B	Review subtracting single digit numbers from the numbers in between 1 and 20. (Week 4)

Week 1

Day	Tiger Session		Topic	Goal
Mon	41	42	Subtraction from 15	15 – 1 digit
Tue	43	44	Subtraction from 15	15 – 1 digit
Wed	45	46	Subtraction from 16	16 – 1 digit
Thu	47	48	Subtraction from 16	16 – 1 digit
Fri	49	50	Review	(15, 16) – 1 digit

Week 2

Day	Tiger Session		Topic	Goal
Mon	51	52	Subtraction from 17	17 – 1 digit
Tue	53	54	Subtraction from 17	17 – 1 digit
Wed	55	56	Subtraction from 18	18 – 1 digit
Thu	57	58	Subtraction from 18	18 – 1 digit
Fri	59	60	Review	(17, 18) – 1 digit

Week 3

Day	Tiger Session		Topic	Goal
Mon	61	62	Subtraction from 19	19 – 1 digit
Tue	63	64	Subtraction from 19	19 – 1 digit
Wed	65	66	Subtraction from 20	20 – 1 digit
Thu	67	68	Subtraction from 20	20 – 1 digit
Fri	69	70	Review	(19, 20) – 1 digit

Week 4

Day	Tiger Session		Topic	Goal
Mon	71	72	Review	(1 ~ 20) – (1 ~ 9)
Tue	73	74	Review	(1 ~ 20) – (1 ~ 9)
Wed	75	76	Review	(1 ~ 20) – (1 ~ 9)
Thu	77	78	Review	(1 ~ 20) – (1 ~ 9)
Fri	79	80	Review	(1 ~ 20) – (1 ~ 9)

Week 1

This week's goal is to subtract single digit numbers from 15 or 16.

Tiger Session

Monday	41	42
Tuesday	43	44
Wednesday	45	46
Thursday	47	48
Friday	49	50

41

Subtraction from 15 ①

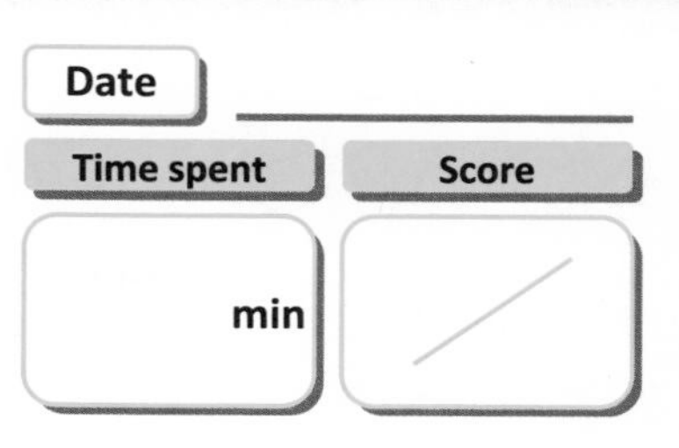

♠ **Subtract.**

1) $15 - 5 =$

2) $15 - 2 =$

3) $15 - 0 =$

4) $15 - 3 =$

5) $15 - 4 =$

6) $10 - 6 =$

7) $15 - 6 =$

8) $10 - 9 =$

9) $15 - 9 =$

10) $10 - 8 =$

11) $15 - 8 =$

12) $10 - 7 =$

13) $15 - 7 =$

14) $10 - 9 =$

15) $15 - 9 =$

16) $10 - 6 =$

17) $15 - 6 =$

18) $15 - 5 =$

19) $15 - 8 =$

20) $15 - 7 =$

Date

Time spent | Score

min

42 Subtraction from 15 ②

♠ Subtract.

1) $15 - 5 =$

2) $15 - 7 =$

3) $15 - 1 =$

4) $15 - 3 =$

5) $15 - 2 =$

6) $15 - 1 =$

7) $15 - 6 =$

8) $15 - 9 =$

9) $15 - 8 =$

10) There were 15 boats on a lake, but after a while, 8 boats leave. How many boats are still on the lake?

Equation: ______________________________

Answer: ______ boats

11) There are some boats on a lake. 15 boats are red, and 6 boats are blue. How many more red boats are there?

Equation: ______________________________

Answer: ______ red boats

43

Subtraction from 15 ③

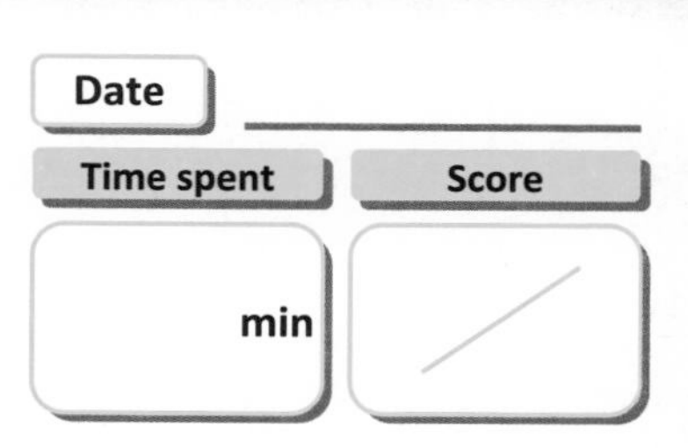

♠ **Subtract.**

1) $\begin{array}{r} 15 \\ -\ \ 5 \\ \hline \end{array}$

2) $\begin{array}{r} 15 \\ -\ \ 6 \\ \hline \end{array}$

3) $\begin{array}{r} 10 \\ -\ \ 8 \\ \hline \end{array}$

4) $\begin{array}{r} 15 \\ -\ \ 8 \\ \hline \end{array}$

5) $\begin{array}{r} 10 \\ -\ \ 7 \\ \hline \end{array}$

6) $\begin{array}{r} 15 \\ -\ \ 7 \\ \hline \end{array}$

7) $\begin{array}{r} 10 \\ -\ \ 9 \\ \hline \end{array}$

8) $\begin{array}{r} 15 \\ -\ \ 9 \\ \hline \end{array}$

9) $\begin{array}{r} 10 \\ -\ 8 \\ \hline \end{array}$

10) $\begin{array}{r} 15 \\ -\ 8 \\ \hline \end{array}$

11) $\begin{array}{r} 15 \\ -\ 3 \\ \hline \end{array}$

12) $\begin{array}{r} 15 \\ -\ 4 \\ \hline \end{array}$

13) $\begin{array}{r} 10 \\ -\ 9 \\ \hline \end{array}$

14) $\begin{array}{r} 15 \\ -\ 9 \\ \hline \end{array}$

15) $\begin{array}{r} 15 \\ -\ 2 \\ \hline \end{array}$

16) $\begin{array}{r} 10 \\ -\ 7 \\ \hline \end{array}$

17) $\begin{array}{r} 15 \\ -\ 7 \\ \hline \end{array}$

18) $\begin{array}{r} 10 \\ -\ 6 \\ \hline \end{array}$

19) $\begin{array}{r} 15 \\ -\ 6 \\ \hline \end{array}$

20) $\begin{array}{r} 15 \\ -\ 5 \\ \hline \end{array}$

Subtraction from 15 ④

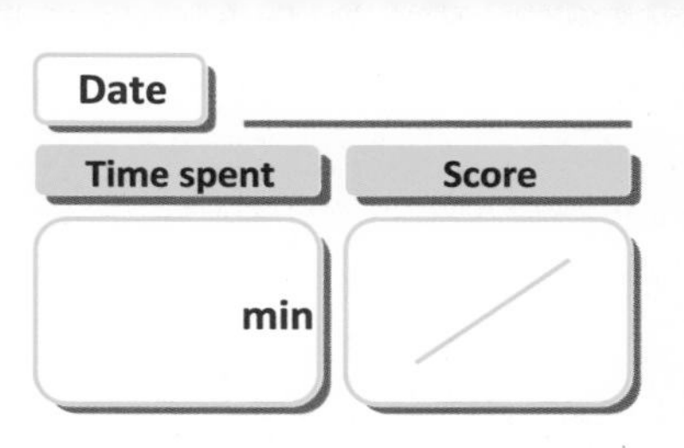

♠ **Subtract.**

1) $\begin{array}{r} 15 \\ -\ 5 \\ \hline \end{array}$

2) $\begin{array}{r} 15 \\ -\ 9 \\ \hline \end{array}$

3) $\begin{array}{r} 15 \\ -\ 7 \\ \hline \end{array}$

4) $\begin{array}{r} 15 \\ -\ 2 \\ \hline \end{array}$

5) $\begin{array}{r} 15 \\ -\ 6 \\ \hline \end{array}$

6) $\begin{array}{r} 15 \\ -\ 3 \\ \hline \end{array}$

7) $\begin{array}{r} 15 \\ -\ 4 \\ \hline \end{array}$

8) $\begin{array}{r} 15 \\ -\ 8 \\ \hline \end{array}$

9) I studied math for 9 minutes yesterday and for 15 minutes today. How many more minutes did I study math today?

Equation: ______________________________

Answer: ______ minutes

10) Today, you drank 7 cups of water, and your mom drank 15 cups. How many more cups did your mom drink?

Equation: ______________________________

Answer: ______ cups

Date

Time spent | Score

min

45

Subtraction from 16 ①

♠ **Subtract.**

1) $16 - 4 =$

2) $16 - 1 =$

3) $16 - 3 =$

4) $16 - 2 =$

5) $16 - 5 =$

6) $16 - 6 =$

7) $16 - 7 =$

8) $16 - 8 =$

9) $16 - 9 =$

10) $10 - 9 =$

11) $16 - 9 =$

12) $10 - 7 =$

13) $16 - 7 =$

14) $10 - 6 =$

15) $16 - 6 =$

16) $10 - 8 =$

17) $16 - 8 =$

18) $16 - 4 =$

19) $16 - 3 =$

20) $16 - 5 =$

46 Subtraction from 16 ②

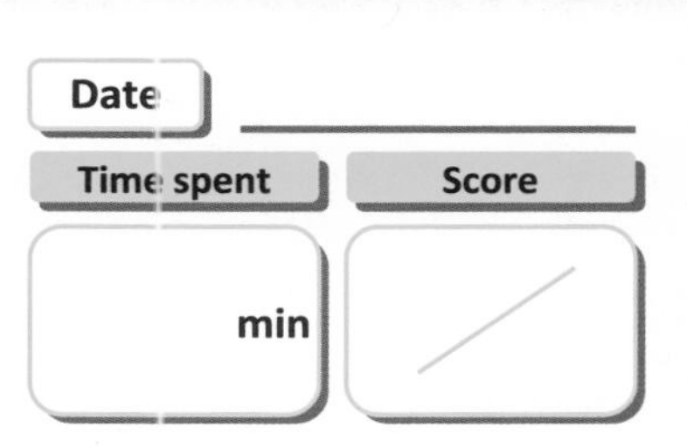

♠ **Subtract.**

1) $16 - 2 =$

2) $16 - 6 =$

3) $16 - 8 =$

4) $16 - 4 =$

5) $16 - 7 =$

6) $16 - 5 =$

7) $16 - 1 =$

8) $16 - 9 =$

9) $16 - 3 =$

10) There were 16 cookies on a table. I ate 9 cookies. How many cookies are left?

Equation: ______________________________

Answer: ______ cookies

11) There are 2 plates with cookies on a table. 16 cookies are on the white plate, and 7 cookies are on black plate. How many more cookies are on the white plate?

Equation: ______________________________

Answer: ______ cookies

Subtraction from 16 ③

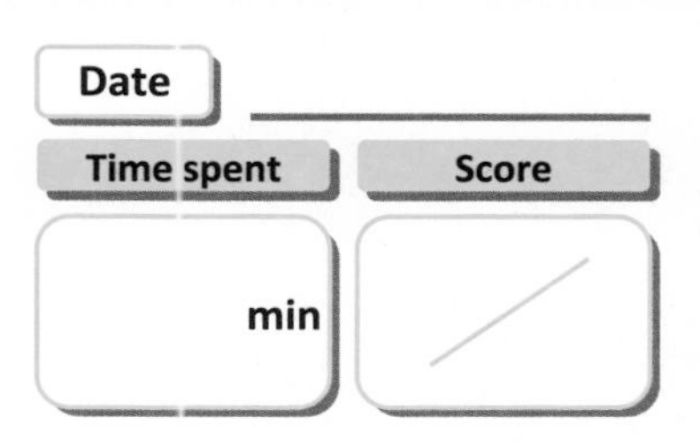

♠ **Subtract.**

1) $10 - 7 =$

2) $16 - 7 =$

3) $10 - 9 =$

4) $16 - 9 =$

5) $10 - 6 =$

6) $16 - 6 =$

7) $10 - 8 =$

8) $16 - 8 =$

9) $\begin{array}{r} 16 \\ -\ 4 \\ \hline \end{array}$

10) $\begin{array}{r} 10 \\ -\ 6 \\ \hline \end{array}$

11) $\begin{array}{r} 16 \\ -\ 6 \\ \hline \end{array}$

12) $\begin{array}{r} 16 \\ -\ 2 \\ \hline \end{array}$

13) $\begin{array}{r} 10 \\ -\ 8 \\ \hline \end{array}$

14) $\begin{array}{r} 16 \\ -\ 8 \\ \hline \end{array}$

15) $\begin{array}{r} 10 \\ -\ 7 \\ \hline \end{array}$

16) $\begin{array}{r} 16 \\ -\ 7 \\ \hline \end{array}$

17) $\begin{array}{r} 16 \\ -\ 1 \\ \hline \end{array}$

18) $\begin{array}{r} 10 \\ -\ 9 \\ \hline \end{array}$

19) $\begin{array}{r} 16 \\ -\ 9 \\ \hline \end{array}$

20) $\begin{array}{r} 16 \\ -\ 3 \\ \hline \end{array}$

48 Subtraction from 16 ④

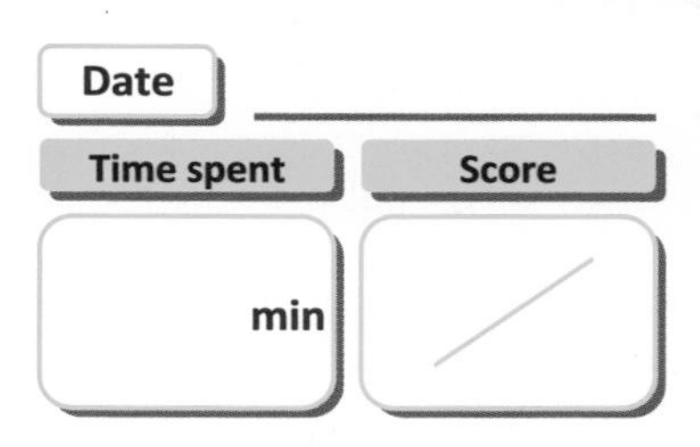

♠ **Subtract.**

1) $\begin{array}{r} 16 \\ -\ 3 \\ \hline \end{array}$

2) $\begin{array}{r} 16 \\ -\ 5 \\ \hline \end{array}$

3) $\begin{array}{r} 16 \\ -\ 7 \\ \hline \end{array}$

4) $\begin{array}{r} 16 \\ -\ 4 \\ \hline \end{array}$

5) $\begin{array}{r} 16 \\ -\ 2 \\ \hline \end{array}$

6) $\begin{array}{r} 16 \\ -\ 9 \\ \hline \end{array}$

7) $\begin{array}{r} 16 \\ -\ 6 \\ \hline \end{array}$

8) $\begin{array}{r} 16 \\ -\ 8 \\ \hline \end{array}$

9) You have 16 pencils. If you give 6 pencils to your sister, how many pencils will you have left?

Equation: ______________________________

Answer: ______ pencils

10) You have 16 pencils and 8 erasers. How many more pencils do you have than erasers?

Equation: ______________________________

Answer: ______ pencils

Date

Time spent | Score

min

49

Review ①
Subtraction from 15, 16

♠ **Subtract.**

1) $10 - 7 =$

2) $15 - 7 =$

3) $16 - 7 =$

4) $15 - 9 =$

5) $16 - 9 =$

6) $\begin{array}{r} 15 \\ -\ \ 8 \\ \hline \end{array}$

7) $\begin{array}{r} 16 \\ -\ \ 8 \\ \hline \end{array}$

8) $\begin{array}{r} 15 \\ -\ \ 6 \\ \hline \end{array}$

9) $\begin{array}{r} 16 \\ -\ \ 6 \\ \hline \end{array}$

10) $15 - 3 =$

11) $16 - 2 =$

12) $15 - 4 =$

13) $16 - 9 =$

14) $15 - 7 =$

15) $\begin{array}{r} 15 \\ -\ \ 8 \\ \hline \end{array}$

16) $\begin{array}{r} 16 \\ -\ \ 3 \\ \hline \end{array}$

17) $\begin{array}{r} 16 \\ -\ \ 7 \\ \hline \end{array}$

18) $\begin{array}{r} 15 \\ -\ \ 9 \\ \hline \end{array}$

19) $\begin{array}{r} 15 \\ -\ \ 3 \\ \hline \end{array}$

20) $\begin{array}{r} 16 \\ -\ \ 1 \\ \hline \end{array}$

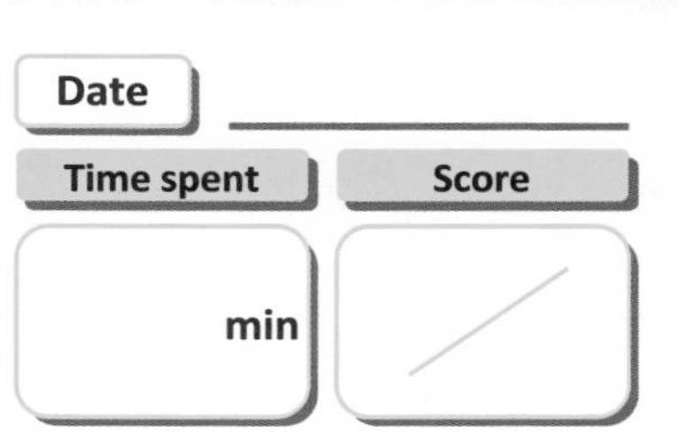

50 Review ②

Subtraction from 15, 16

♠ **Subtract.**

1) $15 - 5 =$

2) $16 - 5 =$

3) $10 - 6 =$

4) $15 - 6 =$

5) $16 - 6 =$

6) $\begin{array}{r} 15 \\ - \ \ 4 \\ \hline \end{array}$

7) $\begin{array}{r} 16 \\ - \ \ 8 \\ \hline \end{array}$

8) $\begin{array}{r} 16 \\ - \ \ 7 \\ \hline \end{array}$

9) $\begin{array}{r} 15 \\ - \ \ 9 \\ \hline \end{array}$

10) In a refrigerator, there were 16 slices of cheese. After I ate 2 slices, how many slices of cheese are left?

Equation: ______________________

Answer: ______ slices

11) In a refrigerator, there are 15 cups of yogurt and 9 slices of cheese. How many more cups of yogurt are there?

Equation: ______________________

Answer: ______ cups

Week 2

This week's goal is to subtract single digit numbers from 17 or 18.

Tiger Session

Monday	51	52
Tuesday	53	54
Wednesday	55	56
Thursday	57	58
Friday	59	60

51 Subtraction from 17 ①

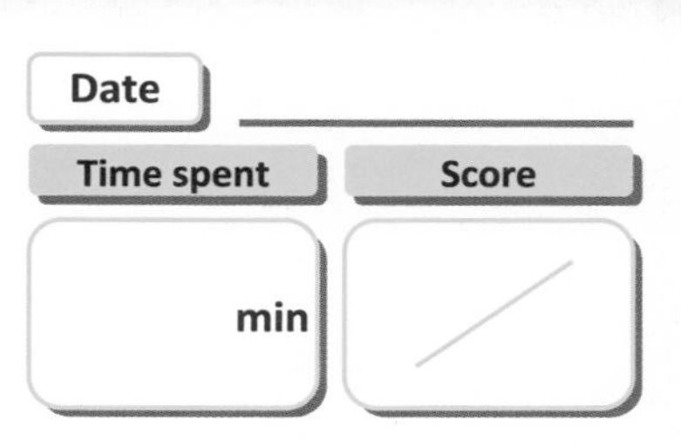

♠ **Subtract.**

1) $17 - 5 =$

2) $17 - 3 =$

3) $17 - 2 =$

4) $17 - 6 =$

5) $17 - 1 =$

6) $17 - 4 =$

7) $17 - 7 =$

8) $17 - 8 =$

9) $17 - 9 =$

10) $10 - 8 =$

11) $17 - 8 =$

12) $10 - 9 =$

13) $17 - 9 =$

14) $10 - 7 =$

15) $17 - 7 =$

16) $17 - 6 =$

17) $17 - 4 =$

18) $17 - 8 =$

19) $17 - 5 =$

20) $17 - 9 =$

52 Subtraction from 17 ②

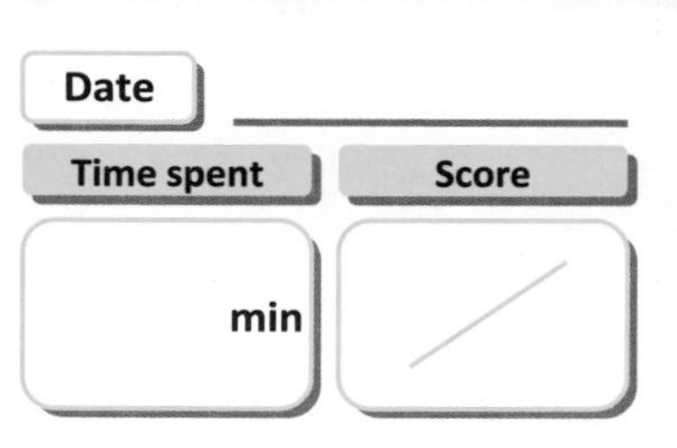

♠ **Subtract.**

1) $17 - 7 =$

2) $17 - 6 =$

3) $17 - 3 =$

4) $17 - 9 =$

5) $17 - 5 =$

6) $17 - 2 =$

7) $17 - 4 =$

8) $17 - 1 =$

9) $17 - 8 =$

10) Yesterday, you made 17 cars with toy blocks. If you give 6 cars to your brother, how many cars will you have left?

Equation: ______________________________

Answer: ______ cars

11) While playing with toy blocks, you made 8 cars and 17 airplanes. How many more airplanes did you make than cars?

Equation: ______________________________

Answer: ______ airplanes

53 Subtraction from 17 ③

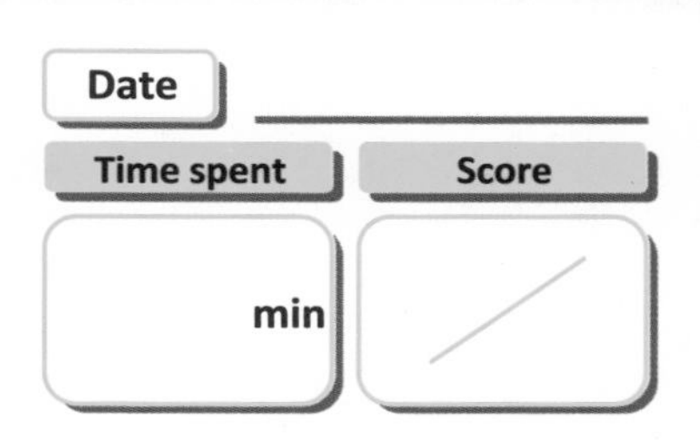

♠ **Subtract.**

1) $\begin{array}{r} 17 \\ -\ 4 \\ \hline \end{array}$

2) $\begin{array}{r} 17 \\ -\ 2 \\ \hline \end{array}$

3) $\begin{array}{r} 10 \\ -\ 8 \\ \hline \end{array}$

4) $\begin{array}{r} 17 \\ -\ 8 \\ \hline \end{array}$

5) $\begin{array}{r} 10 \\ -\ 7 \\ \hline \end{array}$

6) $\begin{array}{r} 17 \\ -\ 7 \\ \hline \end{array}$

7) $\begin{array}{r} 10 \\ -\ 9 \\ \hline \end{array}$

8) $\begin{array}{r} 17 \\ -\ 9 \\ \hline \end{array}$

9) $\begin{array}{r} 10 \\ -\ 9 \\ \hline \end{array}$

10) $\begin{array}{r} 17 \\ -\ 9 \\ \hline \end{array}$

11) $\begin{array}{r} 17 \\ -\ 2 \\ \hline \end{array}$

12) $\begin{array}{r} 10 \\ -\ 7 \\ \hline \end{array}$

13) $\begin{array}{r} 17 \\ -\ 7 \\ \hline \end{array}$

14) $\begin{array}{r} 17 \\ -\ 4 \\ \hline \end{array}$

15) $\begin{array}{r} 17 \\ -\ 5 \\ \hline \end{array}$

16) $\begin{array}{r} 10 \\ -\ 8 \\ \hline \end{array}$

17) $\begin{array}{r} 17 \\ -\ 8 \\ \hline \end{array}$

18) $\begin{array}{r} 17 \\ -\ 9 \\ \hline \end{array}$

19) $\begin{array}{r} 17 \\ -\ 1 \\ \hline \end{array}$

20) $\begin{array}{r} 17 \\ -\ 6 \\ \hline \end{array}$

54 Subtraction from 17 ④

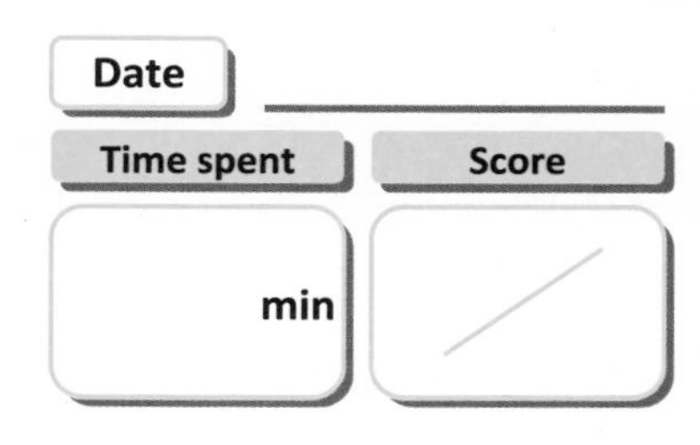

♠ **Subtract.**

1) $\begin{array}{r} 17 \\ -\ \ 3 \\ \hline \end{array}$

2) $\begin{array}{r} 17 \\ -\ \ 6 \\ \hline \end{array}$

3) $\begin{array}{r} 17 \\ -\ \ 4 \\ \hline \end{array}$

4) $\begin{array}{r} 17 \\ -\ \ 7 \\ \hline \end{array}$

5) $\begin{array}{r} 17 \\ -\ \ 8 \\ \hline \end{array}$

6) $\begin{array}{r} 17 \\ -\ \ 2 \\ \hline \end{array}$

7) $\begin{array}{r} 17 \\ -\ \ 9 \\ \hline \end{array}$

8) $\begin{array}{r} 17 \\ -\ \ 5 \\ \hline \end{array}$

9) You helped your mom for 17 minutes this week. If you helped her 7 more minutes this week than last week, how many minutes did you help her last week?

Equation: ______________________

Answer: ______ minutes

10) My family donated 9 pieces of clothing to the Salvation Army last month and 17 pieces of clothing this month. How many more pieces of clothing did my family donate this month?

Equation: ______________________

Answer: ______ pieces

55 Subtraction from 18 ①

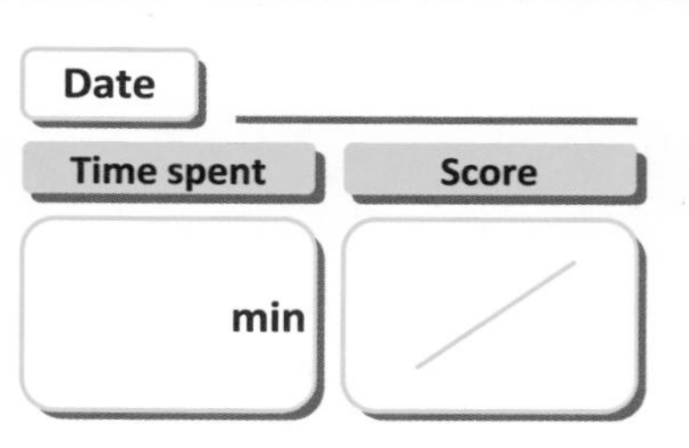

♠ **Subtract.**

1) $18 - 3 =$

2) $18 - 6 =$

3) $18 - 2 =$

4) $18 - 7 =$

5) $18 - 4 =$

6) $18 - 1 =$

7) $18 - 5 =$

8) $18 - 8 =$

9) $18 - 9 =$

10) $10 - 2 =$

11) $15 - 2 =$

12) $18 - 2 =$

13) $8 - 4 =$

14) $18 - 4 =$

15) $10 - 5 =$

16) $15 - 5 =$

17) $18 - 5 =$

18) $8 - 7 =$

19) $18 - 7 =$

20) $18 - 9 =$

56 Subtraction from 18 ②

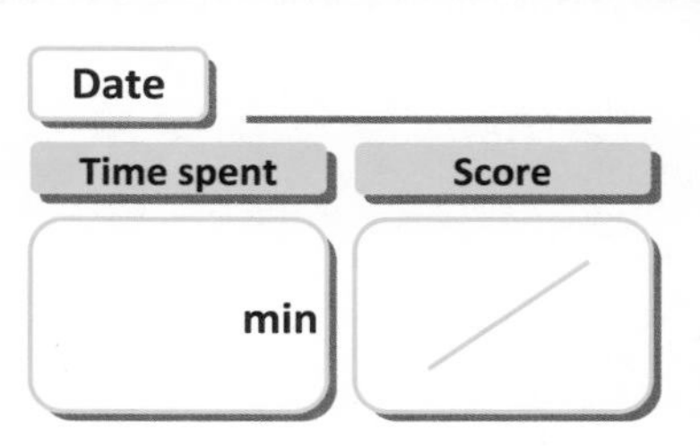

♠ **Subtract.**

1) $18 - 3 =$

2) $18 - 2 =$

3) $18 - 1 =$

4) $18 - 5 =$

5) $18 - 4 =$

6) $18 - 9 =$

7) $18 - 6 =$

8) $18 - 8 =$

9) $18 - 7 =$

10) There were 18 kids on the playground. After 6 leave, how many kids are left on the playground?

Equation: ______________________________

Answer: ______ kids

11) Some children are playing on the playground. 18 are playing in the sandbox, and 7 are playing on the slide. How many more children are playing in the sandbox than on the slide?

Equation: ______________________________

Answer: ______ children

Date

Time spent | Score

min

57 Subtraction from 18 ③

♠ **Subtract.**

1) $\begin{array}{r} 10 \\ -\ \ 3 \\ \hline \end{array}$

2) $\begin{array}{r} 18 \\ -\ \ 3 \\ \hline \end{array}$

3) $\begin{array}{r} 8 \\ -\ 5 \\ \hline \end{array}$

4) $\begin{array}{r} 18 \\ -\ \ 5 \\ \hline \end{array}$

5) $\begin{array}{r} 8 \\ -\ 4 \\ \hline \end{array}$

6) $\begin{array}{r} 18 \\ -\ \ 4 \\ \hline \end{array}$

7) $\begin{array}{r} 10 \\ -\ \ 9 \\ \hline \end{array}$

8) $\begin{array}{r} 18 \\ -\ \ 9 \\ \hline \end{array}$

9) $\begin{array}{r} 18 \\ -\ 8 \\ \hline \end{array}$

10) $\begin{array}{r} 8 \\ -\ 3 \\ \hline \end{array}$

11) $\begin{array}{r} 18 \\ -\ 3 \\ \hline \end{array}$

12) $\begin{array}{r} 18 \\ -\ 6 \\ \hline \end{array}$

13) $\begin{array}{r} 8 \\ -\ 2 \\ \hline \end{array}$

14) $\begin{array}{r} 18 \\ -\ 2 \\ \hline \end{array}$

15) $\begin{array}{r} 18 \\ -\ 4 \\ \hline \end{array}$

16) $\begin{array}{r} 8 \\ -\ 5 \\ \hline \end{array}$

17) $\begin{array}{r} 18 \\ -\ 5 \\ \hline \end{array}$

18) $\begin{array}{r} 18 \\ -\ 8 \\ \hline \end{array}$

19) $\begin{array}{r} 8 \\ -\ 7 \\ \hline \end{array}$

20) $\begin{array}{r} 18 \\ -\ 7 \\ \hline \end{array}$

58 Subtraction from 18 ④

Date

Time spent: min

Score

♠ **Subtract.**

1) $\begin{array}{r} 18 \\ -\ \ 3 \\ \hline \end{array}$

2) $\begin{array}{r} 18 \\ -\ \ 8 \\ \hline \end{array}$

3) $\begin{array}{r} 18 \\ -\ \ 5 \\ \hline \end{array}$

4) $\begin{array}{r} 18 \\ -\ \ 6 \\ \hline \end{array}$

5) $\begin{array}{r} 18 \\ -\ \ 7 \\ \hline \end{array}$

6) $\begin{array}{r} 18 \\ -\ \ 4 \\ \hline \end{array}$

7) $\begin{array}{r} 18 \\ -\ \ 9 \\ \hline \end{array}$

8) $\begin{array}{r} 18 \\ -\ \ 2 \\ \hline \end{array}$

9) Maggie read a book for 18 minutes, and Brian read for 9 minutes. How many more minutes did Maggie read than Brian?

Equation: ______________________________

Answer: ______ minutes

10) Lena decides to read a book for 18 minutes today. If she read for 8 minutes, how many more minutes does she have to read to reach her goal?

Equation: ______________________________

Answer: ______ minutes

Date

Time spent | Score

min

59 Review ①
Subtraction from 17, 18

♠ **Subtract.**

1) $17 - 5 =$

2) $18 - 5 =$

3) $10 - 9 =$

4) $17 - 9 =$

5) $18 - 9 =$

6) $\begin{array}{r} 17 \\ -\ \ 6 \\ \hline \end{array}$

7) $\begin{array}{r} 18 \\ -\ \ 6 \\ \hline \end{array}$

8) $\begin{array}{r} 17 \\ -\ \ 3 \\ \hline \end{array}$

9) $\begin{array}{r} 18 \\ -\ \ 3 \\ \hline \end{array}$

10) $18 - 6 =$

11) $7 - 6 =$

12) $17 - 6 =$

13) $18 - 4 =$

14) $17 - 4 =$

15) $\begin{array}{r} 18 \\ -\ \ 2 \\ \hline \end{array}$

16) $\begin{array}{r} 17 \\ -\ \ 2 \\ \hline \end{array}$

17) $\begin{array}{r} 17 \\ -\ \ 7 \\ \hline \end{array}$

18) $\begin{array}{r} 18 \\ -\ \ 7 \\ \hline \end{array}$

19) $\begin{array}{r} 18 \\ -\ \ 9 \\ \hline \end{array}$

20) $\begin{array}{r} 17 \\ -\ \ 9 \\ \hline \end{array}$

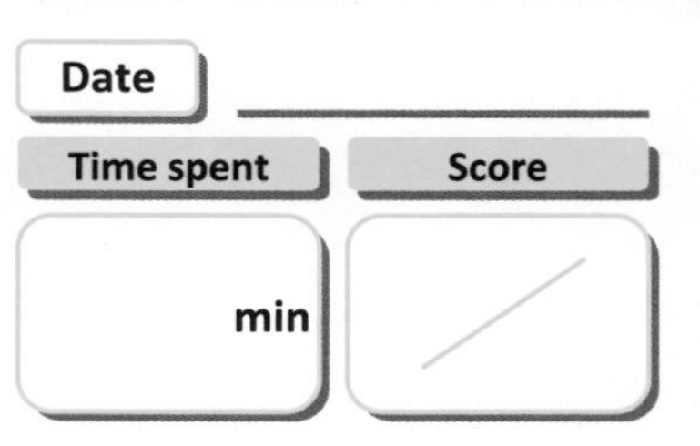

Review ② Subtraction from 17, 18

♠ **Subtract.**

1) $17 - 5 =$

2) $18 - 6 =$

3) $17 - 3 =$

4) $10 - 4 =$

5) $18 - 8 =$

6) $\begin{array}{r} 17 \\ -\ \ 3 \\ \hline \end{array}$

7) $\begin{array}{r} 18 \\ -\ \ 5 \\ \hline \end{array}$

8) $\begin{array}{r} 18 \\ -\ \ 9 \\ \hline \end{array}$

9) $\begin{array}{r} 17 \\ -\ \ 2 \\ \hline \end{array}$

10) In your class, there are 17 students. If 8 students are girls, how many boys are in your class?

Equation: ______________________

Answer: ______ boys

11) In science class, 18 students planted a seed individually to see how quickly the seeds will sprout. After 3 days, 7 seeds started to sprout. How many seeds have yet to sprout?

Equation: ______________________

Answer: ______ seeds

Week 3

This week's goal is to subtract single digit numbers from 19 or 20.

Tiger Session

Monday	61	62
Tuesday	63	64
Wednesday	65	66
Thursday	67	68
Friday	69	70

Date

Time spent | Score

min

61 Subtraction from 19 ①

♠ **Subtract.**

1) $19 - 1 =$

2) $15 - 3 =$

3) $19 - 3 =$

4) $10 - 8 =$

5) $15 - 8 =$

6) $19 - 8 =$

7) $9 - 6 =$

8) $19 - 6 =$

9) $19 - 9 =$

10) $9 - 7 =$

11) $19 - 7 =$

12) $15 - 4 =$

13) $19 - 4 =$

14) $10 - 4 =$

15) $19 - 4 =$

16) $9 - 2 =$

17) $19 - 2 =$

18) $10 - 5 =$

19) $15 - 5 =$

20) $19 - 5 =$

62

Subtraction from 19 ②

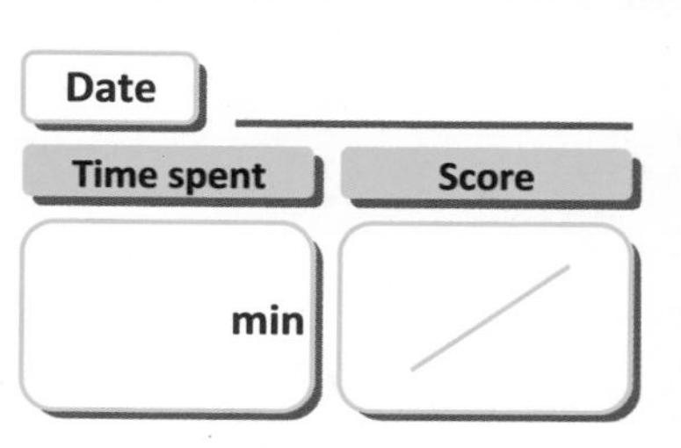

♠ **Subtract.**

1) $19 - 3 =$

2) $19 - 9 =$

3) $19 - 2 =$

4) $19 - 1 =$

5) $19 - 8 =$

6) $19 - 5 =$

7) $19 - 7 =$

8) $19 - 4 =$

9) $19 - 6 =$

10) There were 19 soccer balls in a gym. If some kids took 5, how many soccer balls are left?

Equation: ______________________________

Answer: ______ soccer balls

11) It's lunch time at school. In the cafeteria, 8 students choose chocolate milk, and 19 students choose white milk. How many more students choose white milk?

Equation: ______________________________

Answer: ______ students

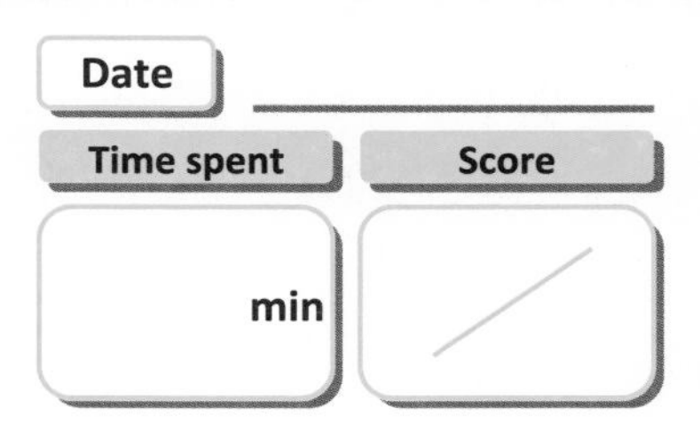

63 Subtraction from 19 ③

♠ **Subtract.**

1) $\begin{array}{r} 19 \\ -\ \ 4 \\ \hline \end{array}$

2) $\begin{array}{r} 19 \\ -\ \ 3 \\ \hline \end{array}$

3) $\begin{array}{r} 19 \\ -\ \ 6 \\ \hline \end{array}$

4) $\begin{array}{r} 19 \\ -\ \ 8 \\ \hline \end{array}$

5) $\begin{array}{r} 19 \\ -\ \ 9 \\ \hline \end{array}$

6) $\begin{array}{r} 19 \\ -\ \ 7 \\ \hline \end{array}$

7) $\begin{array}{r} 19 \\ -\ \ 2 \\ \hline \end{array}$

8) $\begin{array}{r} 19 \\ -\ \ 5 \\ \hline \end{array}$

9) $\begin{array}{r} 9 \\ -\ 4 \\ \hline \end{array}$

10) $\begin{array}{r} 19 \\ -\ 4 \\ \hline \end{array}$

11) $\begin{array}{r} 19 \\ -\ 8 \\ \hline \end{array}$

12) $\begin{array}{r} 19 \\ -\ 1 \\ \hline \end{array}$

13) $\begin{array}{r} 15 \\ -\ 5 \\ \hline \end{array}$

14) $\begin{array}{r} 19 \\ -\ 5 \\ \hline \end{array}$

15) $\begin{array}{r} 10 \\ -\ 7 \\ \hline \end{array}$

16) $\begin{array}{r} 15 \\ -\ 7 \\ \hline \end{array}$

17) $\begin{array}{r} 19 \\ -\ 7 \\ \hline \end{array}$

18) $\begin{array}{r} 19 \\ -\ 3 \\ \hline \end{array}$

19) $\begin{array}{r} 10 \\ -\ 2 \\ \hline \end{array}$

20) $\begin{array}{r} 19 \\ -\ 2 \\ \hline \end{array}$

64 Subtraction from 19 ④

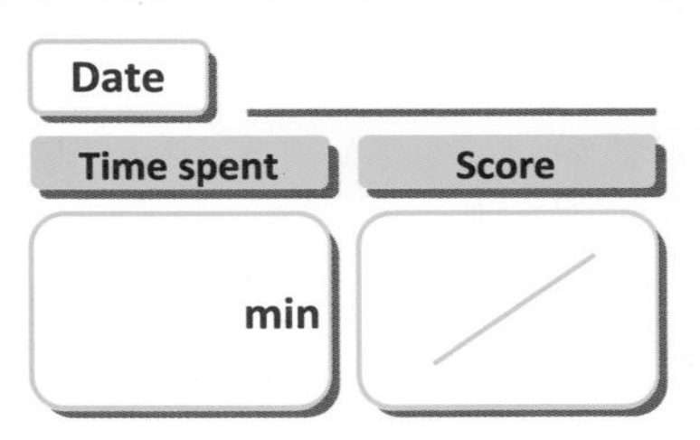

♠ **Subtract.**

1) $\begin{array}{r} 19 \\ -\ 2 \\ \hline \end{array}$

2) $\begin{array}{r} 19 \\ -\ 7 \\ \hline \end{array}$

3) $\begin{array}{r} 19 \\ -\ 9 \\ \hline \end{array}$

4) $\begin{array}{r} 19 \\ -\ 4 \\ \hline \end{array}$

5) $\begin{array}{r} 19 \\ -\ 5 \\ \hline \end{array}$

6) $\begin{array}{r} 19 \\ -\ 3 \\ \hline \end{array}$

7) $\begin{array}{r} 19 \\ -\ 6 \\ \hline \end{array}$

8) $\begin{array}{r} 19 \\ -\ 8 \\ \hline \end{array}$

9) There are 19 students in my class. Today, when the teacher asked a question, 7 students raised their hand. How many students didn't raise their hand?

Equation: ______________________________

Answer: ______ students

10) My teacher had 19 paper clips. After using 6, how many paper clips does she have left?

Equation: ______________________________

Answer: ______ paper clips

Date

Time spent | Score

min

65 Subtraction from 20 ①

♠ **Subtract.**

1) $19 - 1 =$

2) $20 - 1 =$

3) $10 - 2 =$

4) $20 - 2 =$

5) $10 - 4 =$

6) $20 - 4 =$

7) $10 - 5 =$

8) $15 - 5 =$

9) $20 - 5 =$

10) $10 - 3 =$

11) $20 - 3 =$

12) $10 - 8 =$

13) $20 - 8 =$

14) $10 - 7 =$

15) $20 - 7 =$

16) $10 - 6 =$

17) $20 - 6 =$

18) $10 - 9 =$

19) $20 - 9 =$

20) $20 - 4 =$

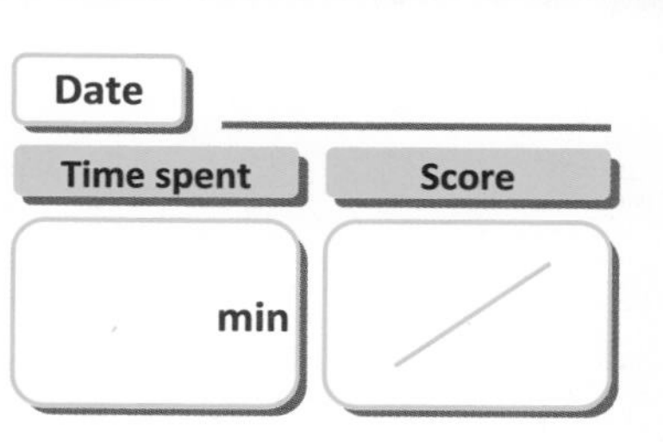

66 Subtraction from 20 ②

♠ **Subtract.**

1) $20 - 6 =$

2) $20 - 2 =$

3) $20 - 1 =$

4) $20 - 9 =$

5) $20 - 4 =$

6) $20 - 7 =$

7) $20 - 3 =$

8) $20 - 8 =$

9) $20 - 5 =$

10) In Joy's class, 20 students come to school on the school bus and 5 students by foot. How many more students used the school bus?

Equation: ______________________

Answer: ______ students

11) Andrew walked to school for 20 days this month. Last month he walked to school for only 7 days. How many more days did he walk to school this month?

Equation: ______________________

Answer: ______ days

67

Subtraction from 20 ③

Date

Time spent: min

Score

♠ **Subtract.**

1) $10 - 1 =$

2) $20 - 1 =$

3) $10 - 3 =$

4) $20 - 3 =$

5) $10 - 5 =$

6) $20 - 5 =$

7) $10 - 7 =$

8) $20 - 7 =$

9) $\begin{array}{r} 10 \\ -\ 8 \\ \hline \end{array}$

10) $\begin{array}{r} 20 \\ -\ 8 \\ \hline \end{array}$

11) $\begin{array}{r} 20 \\ -\ 3 \\ \hline \end{array}$

12) $\begin{array}{r} 10 \\ -\ 9 \\ \hline \end{array}$

13) $\begin{array}{r} 20 \\ -\ 9 \\ \hline \end{array}$

14) $\begin{array}{r} 20 \\ -\ 5 \\ \hline \end{array}$

15) $\begin{array}{r} 20 \\ -\ 1 \\ \hline \end{array}$

16) $\begin{array}{r} 10 \\ -\ 6 \\ \hline \end{array}$

17) $\begin{array}{r} 20 \\ -\ 6 \\ \hline \end{array}$

18) $\begin{array}{r} 20 \\ -\ 2 \\ \hline \end{array}$

19) $\begin{array}{r} 10 \\ -\ 7 \\ \hline \end{array}$

20) $\begin{array}{r} 20 \\ -\ 7 \\ \hline \end{array}$

68

Subtraction from 20 ④

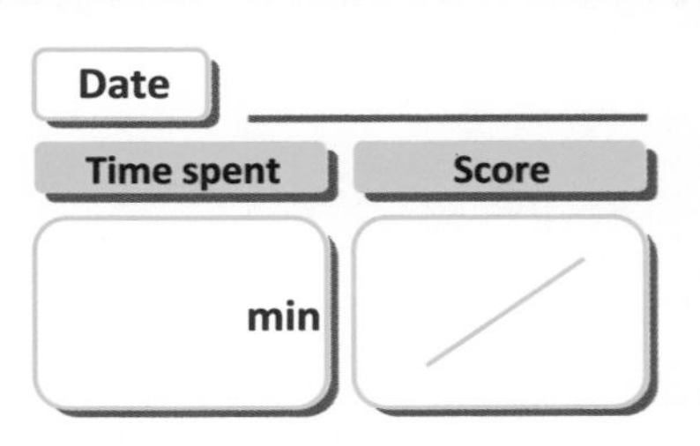

♠ **Subtract.**

1) $\begin{array}{r} 20 \\ -\ \ 4 \\ \hline \end{array}$

2) $\begin{array}{r} 20 \\ -\ \ 3 \\ \hline \end{array}$

3) $\begin{array}{r} 20 \\ -\ \ 8 \\ \hline \end{array}$

4) $\begin{array}{r} 20 \\ -\ \ 7 \\ \hline \end{array}$

5) $\begin{array}{r} 20 \\ -\ \ 2 \\ \hline \end{array}$

6) $\begin{array}{r} 20 \\ -\ \ 5 \\ \hline \end{array}$

7) $\begin{array}{r} 20 \\ -\ \ 9 \\ \hline \end{array}$

8) $\begin{array}{r} 20 \\ -\ \ 6 \\ \hline \end{array}$

9) There were 20 packs of orange juice in a store. After 8 packs are sold, how many packs of orange juice are still left?

Equation: ______________________________

Answer: ______ packs

10) I walked with my family around the neighborhood for 9 minutes yesterday and for 20 minutes today. How many more minutes did we walk today?

Equation: ______________________________

Answer: ______ minutes

Review ①
Subtraction from 19, 20

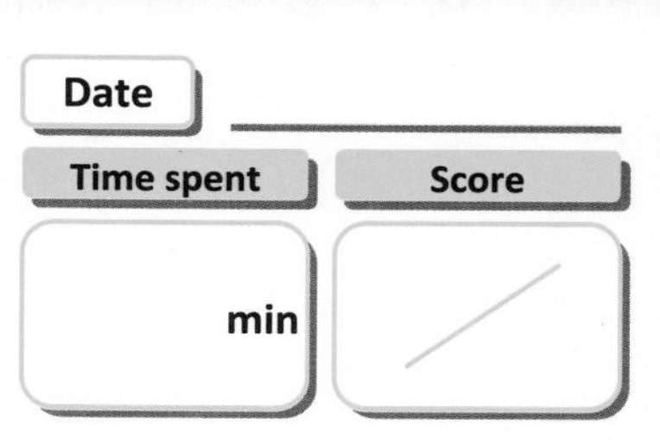

♠ **Subtract.**

1) $10 - 2 =$

2) $19 - 2 =$

3) $20 - 2 =$

4) $19 - 7 =$

5) $20 - 7 =$

6) $\begin{array}{r} 19 \\ -\ \ 5 \\ \hline \end{array}$

7) $\begin{array}{r} 20 \\ -\ \ 5 \\ \hline \end{array}$

8) $\begin{array}{r} 19 \\ -\ \ 3 \\ \hline \end{array}$

9) $\begin{array}{r} 20 \\ -\ \ 3 \\ \hline \end{array}$

10) $10 - 4 =$

11) $19 - 4 =$

12) $20 - 4 =$

13) $19 - 6 =$

14) $20 - 6 =$

15) $\begin{array}{r} 10 \\ -\ \ 1 \\ \hline \end{array}$

16) $\begin{array}{r} 20 \\ -\ \ 1 \\ \hline \end{array}$

17) $\begin{array}{r} 19 \\ -\ \ 3 \\ \hline \end{array}$

18) $\begin{array}{r} 20 \\ -\ \ 8 \\ \hline \end{array}$

19) $\begin{array}{r} 19 \\ -\ \ 9 \\ \hline \end{array}$

20) $\begin{array}{r} 20 \\ -\ \ 9 \\ \hline \end{array}$

70

Review ②
Subtraction from 19, 20

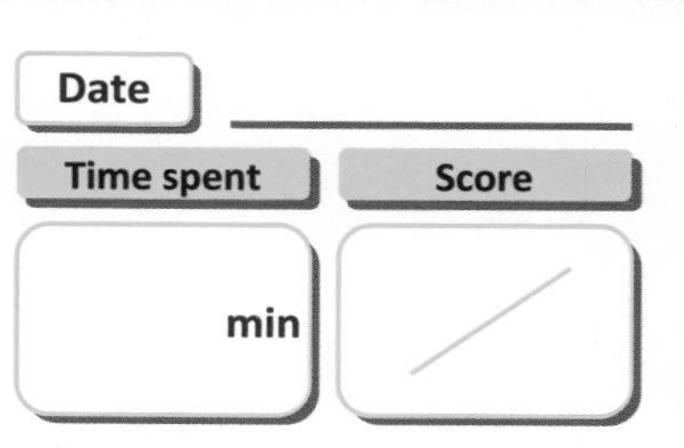

♠ **Subtract.**

1) $10 - 5 =$

2) $19 - 5 =$

3) $20 - 5 =$

4) $10 - 6 =$

5) $20 - 6 =$

6) $\begin{array}{r} 19 \\ -\ \ 7 \\ \hline \end{array}$

7) $\begin{array}{r} 19 \\ -\ \ 9 \\ \hline \end{array}$

8) $\begin{array}{r} 20 \\ -\ \ 3 \\ \hline \end{array}$

9) $\begin{array}{r} 20 \\ -\ \ 9 \\ \hline \end{array}$

10) There are 19 kids live on Stadium Avenue. Among those 19, if 5 kids are boys, how many are girls?

Equation: ______________________________

Answer: ______ girls

11) Mia ran a lemonade stand this morning. After she made 20 cups of lemonade, 6 cups were sold. How many cups are left?

Equation: ______________________________

Answer: ______ cups

Week 4

This week's goal is to subtract single digit numbers from numbers between 1 and 20.

Tiger Session

Monday	71	72
Tuesday	73	74
Wednesday	75	76
Thursday	77	78
Friday	79	80

Date ______

Time spent	Score
min	/

71

Review ①

Subtraction from 1 ~ 20

♠ **Subtract.**

1) $13 - 3 =$

2) $15 - 3 =$

3) $5 - 2 =$

4) $19 - 7 =$

5) $14 - 7 =$

6) $\begin{array}{r} 7 \\ -\ 4 \\ \hline \end{array}$

7) $\begin{array}{r} 10 \\ -\ 3 \\ \hline \end{array}$

8) $\begin{array}{r} 11 \\ -\ 3 \\ \hline \end{array}$

9) $\begin{array}{r} 16 \\ -\ 7 \\ \hline \end{array}$

10) $15 - 5 =$

11) $15 - 7 =$

12) $4 - 1 =$

13) $12 - 4 =$

14) $6 - 3 =$

15) $\begin{array}{r} 18 \\ -\ \ 6 \\ \hline \end{array}$

16) $\begin{array}{r} 20 \\ -\ \ 6 \\ \hline \end{array}$

17) $\begin{array}{r} 14 \\ -\ \ 5 \\ \hline \end{array}$

18) $\begin{array}{r} 9 \\ -\ \ 3 \\ \hline \end{array}$

19) $\begin{array}{r} 12 \\ -\ \ 6 \\ \hline \end{array}$

20) $\begin{array}{r} 10 \\ -\ \ 8 \\ \hline \end{array}$

72

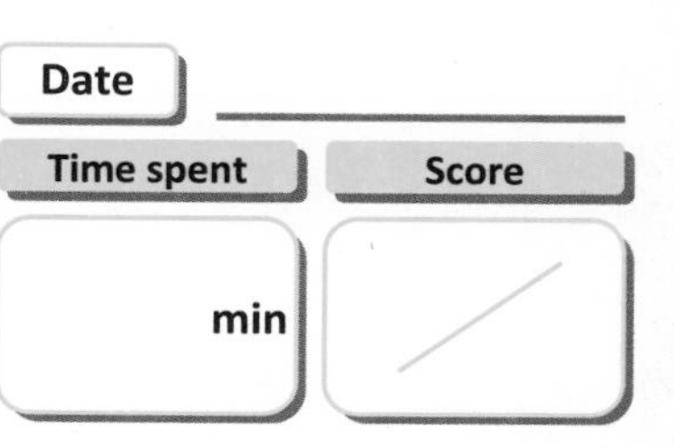

Review ②
Subtraction from 1 ~ 20

♠ **Subtract.**

1) $7 - 3 =$

2) $13 - 6 =$

3) $19 - 4 =$

4) $15 - 5 =$

5) $9 - 5 =$

6) $\begin{array}{r} 10 \\ -\ 7 \\ \hline \end{array}$

7) $\begin{array}{r} 11 \\ -\ 7 \\ \hline \end{array}$

8) $\begin{array}{r} 17 \\ -\ 4 \\ \hline \end{array}$

9) $\begin{array}{r} 13 \\ -\ 7 \\ \hline \end{array}$

10) There were 12 children playing in the gym. After a while, if only 7 children are in the gym, how many children left the gym?

Equation: ______________________________

Answer: ______ children

11) In a grocery store, 14 people were in line to check out. After a while, 5 people finished checking out and left the store. If no one else joined the line to check out, how many people are still left in line?

Equation: ______________________________

Answer: ______ people

Date ____

Time spent | min

Score

73 Review ③

Subtraction from 1 ~ 20

♠ **Subtract.**

1) $10 - 8 =$

2) $15 - 8 =$

3) $20 - 8 =$

4) $11 - 5 =$

5) $13 - 5 =$

6) $9 - 4$

7) $15 - 6$

8) $19 - 4$

9) $13 - 6$

10) $10 - 3 =$

11) $6 - 4 =$

12) $8 - 5 =$

13) $10 - 7 =$

14) $12 - 5 =$

15) $\begin{array}{r} 14 \\ -\ \ 5 \\ \hline \end{array}$

16) $\begin{array}{r} 18 \\ -\ \ 8 \\ \hline \end{array}$

17) $\begin{array}{r} 14 \\ -\ \ 2 \\ \hline \end{array}$

18) $\begin{array}{r} 18 \\ -\ \ 3 \\ \hline \end{array}$

19) $\begin{array}{r} 17 \\ -\ \ 1 \\ \hline \end{array}$

20) $\begin{array}{r} 19 \\ -\ \ 8 \\ \hline \end{array}$

74

Review ④
Subtraction from 1 ~ 20

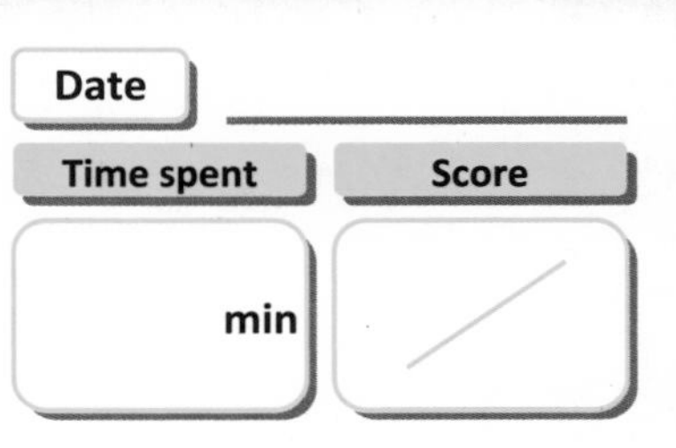

♠ **Subtract.**

1) $8 - 2 =$

2) $18 - 3 =$

3) $10 - 4 =$

4) $17 - 5 =$

5) $13 - 6 =$

6) $\begin{array}{r} 11 \\ -\ \ 7 \\ \hline \end{array}$

7) $\begin{array}{r} 15 \\ -\ \ 8 \\ \hline \end{array}$

8) $\begin{array}{r} 18 \\ -\ \ 9 \\ \hline \end{array}$

9) $\begin{array}{r} 12 \\ -\ \ 9 \\ \hline \end{array}$

10) There are 20 birds in a tree. After 8 birds leave, how many birds are left in the tree?

Equation: ______________________________

Answer: ______ birds

11) Mom bought 15 packs of fruit punch at the store and put them in the refrigerator. The kids drank 7 packs of fruit punch. How many packs of fruit punch are left?

Equation: ______________________________

Answer: ______ packs

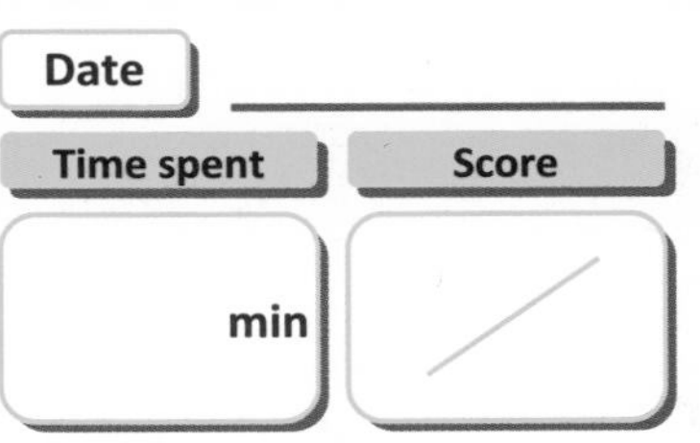

75

Review ⑤
Subtraction from 1 ~ 20

♠ **Subtract.**

1) $3 - 1 =$

2) $13 - 2 =$

3) $18 - 3 =$

4) $17 - 4 =$

5) $9 - 5 =$

6) $\begin{array}{r} 12 \\ -\ \ 6 \\ \hline \end{array}$

7) $\begin{array}{r} 12 \\ -\ \ 8 \\ \hline \end{array}$

8) $\begin{array}{r} 16 \\ -\ \ 7 \\ \hline \end{array}$

9) $\begin{array}{r} 15 \\ -\ \ 9 \\ \hline \end{array}$

10) $8 - 5 =$

11) $5 - 4 =$

12) $15 - 3 =$

13) $20 - 2 =$

14) $19 - 6 =$

15) $\begin{array}{r} 14 \\ - \;\; 8 \\ \hline \end{array}$

16) $\begin{array}{r} 18 \\ - \;\; 8 \\ \hline \end{array}$

17) $\begin{array}{r} 17 \\ - \;\; 8 \\ \hline \end{array}$

18) $\begin{array}{r} 12 \\ - \;\; 4 \\ \hline \end{array}$

19) $\begin{array}{r} 19 \\ - \;\; 9 \\ \hline \end{array}$

20) $\begin{array}{r} 16 \\ - \;\; 7 \\ \hline \end{array}$

76

Review ⑥

Subtraction from 1 ~ 20

Date ______

Time spent: ____ min | Score: ____

♠ **Subtract.**

1) $2 - 2 =$

2) $4 - 2 =$

3) $6 - 2 =$

4) $8 - 4 =$

5) $10 - 4 =$

6) $12 - 6$

7) $14 - 6$

8) $16 - 8$

9) $18 - 8$

10) In the garden, there are 18 pink flowers and 7 yellow flowers. How many more pink flowers are in the garden than yellow ones?

Equation: ______________________________

Answer: ______ pink flowers

11) There were 15 frogs playing in the pond. After a while, 7 frogs leave. How many frogs are still in the pond?

Equation: ______________________________

Answer: ______ frogs

Date

Time spent	Score
min	

77 Review ⑦

Subtraction from 1 ~ 20

♠ **Subtract.**

1) $12 - 3 =$

2) $8 - 3 =$

3) $10 - 9 =$

4) $15 - 9 =$

5) $13 - 9 =$

6) $\begin{array}{r} 19 \\ -\ \ 8 \\ \hline \end{array}$

7) $\begin{array}{r} 11 \\ -\ \ 8 \\ \hline \end{array}$

8) $\begin{array}{r} 10 \\ -\ \ 7 \\ \hline \end{array}$

9) $\begin{array}{r} 16 \\ -\ \ 7 \\ \hline \end{array}$

10) $9 - 6 =$

11) $16 - 6 =$

12) $8 - 5 =$

13) $13 - 5 =$

14) $10 - 3 =$

15) $\begin{array}{r} 16 \\ -\ \ 4 \\ \hline \end{array}$

16) $\begin{array}{r} 11 \\ -\ \ 4 \\ \hline \end{array}$

17) $\begin{array}{r} 18 \\ -\ \ 5 \\ \hline \end{array}$

18) $\begin{array}{r} 14 \\ -\ \ 5 \\ \hline \end{array}$

19) $\begin{array}{r} 12 \\ -\ \ 6 \\ \hline \end{array}$

20) $\begin{array}{r} 7 \\ -\ \ 4 \\ \hline \end{array}$

78

Review ⑧
Subtraction from 1 ~ 20

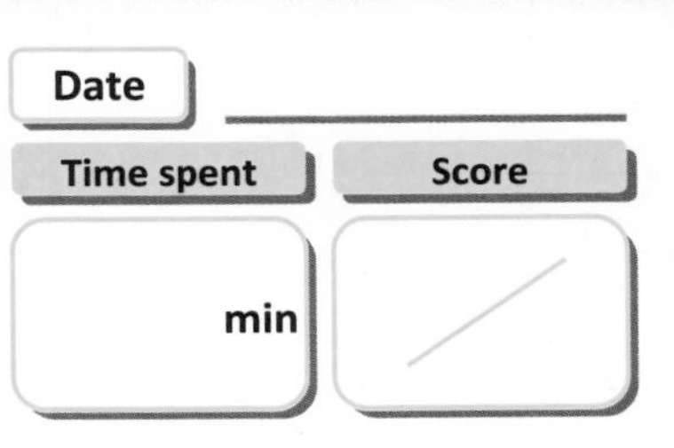

♠ **Subtract.**

1) $3 - 1 =$

2) $6 - 5 =$

3) $9 - 4 =$

4) $12 - 4 =$

5) $15 - 3 =$

6) $\begin{array}{r} 18 \\ -\ \ 5 \\ \hline \end{array}$

7) $\begin{array}{r} 16 \\ -\ \ 3 \\ \hline \end{array}$

8) $\begin{array}{r} 14 \\ -\ \ 8 \\ \hline \end{array}$

9) $\begin{array}{r} 12 \\ -\ \ 9 \\ \hline \end{array}$

10) In a pride of lions, there are 16 adult lions and 3 baby lions. How many more adult lions are there than baby lions?

Equation: ______________________________

Answer: ______ adult lions

11) In a forest, 17 monkeys were playing. After some time, 5 monkeys leave to find some food. How many monkeys are still playing?

Equation: ______________________________

Answer: ______ monkeys

79

Review ⑨
Subtraction from 1 ~ 20

Date ____

Time spent	Score
min	/

♠ **Subtract.**

1) $10 - 2 =$

2) $18 - 2 =$

3) $10 - 8 =$

4) $16 - 8 =$

5) $9 - 6 =$

6) $\begin{array}{r} 15 \\ -\ \ 7 \\ \hline \end{array}$

7) $\begin{array}{r} 14 \\ -\ \ 7 \\ \hline \end{array}$

8) $\begin{array}{r} 13 \\ -\ \ 8 \\ \hline \end{array}$

9) $\begin{array}{r} 19 \\ -\ \ 8 \\ \hline \end{array}$

10) $8 - 5 =$

11) $18 - 5 =$

12) $7 - 6 =$

13) $17 - 9 =$

14) $14 - 2 =$

15) $\begin{array}{r} 11 \\ -\ \ 4 \\ \hline \end{array}$

16) $\begin{array}{r} 15 \\ -\ \ 6 \\ \hline \end{array}$

17) $\begin{array}{r} 13 \\ -\ \ 8 \\ \hline \end{array}$

18) $\begin{array}{r} 19 \\ -\ \ 6 \\ \hline \end{array}$

19) $\begin{array}{r} 14 \\ -\ \ 6 \\ \hline \end{array}$

20) $\begin{array}{r} 12 \\ -\ \ 7 \\ \hline \end{array}$

Date

Time spent | Score

min

80

Review ⑩
Subtraction from 1 ~ 20

♠ **Subtract.**

1) $9 - 7 =$

2) $12 - 7 =$

3) $18 - 6 =$

4) $15 - 6 =$

5) $13 - 5 =$

6) $\begin{array}{r} 10 \\ -\ 8 \\ \hline \end{array}$

7) $\begin{array}{r} 17 \\ -\ 8 \\ \hline \end{array}$

8) $\begin{array}{r} 14 \\ -\ 8 \\ \hline \end{array}$

9) $\begin{array}{r} 10 \\ -\ 6 \\ \hline \end{array}$

10) Christine had 19 cans of food in her pantry. After donating 3, how many cans does she have left?

Equation: ______________________________

Answer: ______ cans

11) Max helped 5 classmates last month and 12 classmates this month. How many more classmates did he help this month?

Equation: ______________________________

Answer: ______ classmates

C – 2: Answers

Week 1

41 (p. 5 ~ 6)

① 10 ② 13 ③ 15 ④ 12 ⑤ 11
⑥ 4 ⑦ 9 ⑧ 1 ⑨ 6 ⑩ 2
⑪ 7 ⑫ 3 ⑬ 8 ⑭ 1 ⑮ 6
⑯ 4 ⑰ 9 ⑱ 10 ⑲ 7 ⑳ 8

42 (p. 7 ~ 8)

① 10 ② 8 ③ 14 ④ 12 ⑤ 13
⑥ 14 ⑦ 9 ⑧ 6 ⑨ 7
⑩ 15 – 8 = 7, 7 ⑪ 15 – 6 = 9, 9

43 (p. 9 ~ 10)

① 10 ② 9 ③ 2 ④ 7 ⑤ 3
⑥ 8 ⑦ 1 ⑧ 6 ⑨ 2 ⑩ 7
⑪ 12 ⑫ 11 ⑬ 1 ⑭ 6 ⑮ 13
⑯ 3 ⑰ 8 ⑱ 4 ⑲ 9 ⑳ 10

44 (p. 11 ~ 12)

① 10 ② 6 ③ 8 ④ 13 ⑤ 9
⑥ 12 ⑦ 11 ⑧ 7
⑨ 15 – 9 = 6, 6 ⑩ 15 – 7 = 8, 8

45 (p. 13 ~ 14)

① 12 ② 15 ③ 13 ④ 14 ⑤ 11
⑥ 10 ⑦ 9 ⑧ 8 ⑨ 7 ⑩ 1
⑪ 7 ⑫ 3 ⑬ 9 ⑭ 4 ⑮ 10
⑯ 2 ⑰ 8 ⑱ 12 ⑲ 13 ⑳ 11

46 (p. 15 ~ 16)

① 14 ② 10 ③ 8 ④ 12 ⑤ 9
⑥ 11 ⑦ 15 ⑧ 7 ⑨ 13
⑩ 16 – 9 = 7, 7 ⑪ 16 – 7 = 9, 9

47 (p. 17 ~ 18)

① 3 ② 9 ③ 1 ④ 7 ⑤ 4
⑥ 10 ⑦ 2 ⑧ 8 ⑨ 12 ⑩ 4
⑪ 10 ⑫ 14 ⑬ 2 ⑭ 8 ⑮ 3
⑯ 9 ⑰ 15 ⑱ 1 ⑲ 7 ⑳ 13

48 (p. 19 ~ 20)

① 13 ② 11 ③ 9 ④ 12 ⑤ 14
⑥ 7 ⑦ 10 ⑧ 8
⑨ 16 – 6 = 10, 10 ⑩ 16 – 8 = 8, 8

49 (p. 21 ~ 22)

① 3 ② 8 ③ 9 ④ 6 ⑤ 7
⑥ 7 ⑦ 8 ⑧ 9 ⑨ 10 ⑩ 12
⑪ 14 ⑫ 11 ⑬ 7 ⑭ 8 ⑮ 7
⑯ 13 ⑰ 9 ⑱ 6 ⑲ 12 ⑳ 15

50 (p. 23 ~ 24)

① 10 ② 11 ③ 4 ④ 9 ⑤ 10
⑥ 11 ⑦ 8 ⑧ 9 ⑨ 6
⑩ 16 – 2 = 14, 14 ⑪ 15 – 9 = 6, 6

Week 2

51 (p. 27 ~ 28)

① 12 ② 14 ③ 15 ④ 11 ⑤ 16
⑥ 13 ⑦ 10 ⑧ 9 ⑨ 8 ⑩ 2
⑪ 9 ⑫ 1 ⑬ 8 ⑭ 3 ⑮ 10
⑯ 11 ⑰ 13 ⑱ 9 ⑲ 12 ⑳ 8

52 (p. 29 ~ 30)

① 10 ② 11 ③ 14 ④ 8 ⑤ 12
⑥ 15 ⑦ 13 ⑧ 16 ⑨ 9
⑩ 17 – 6 = 11, 11 ⑪ 17 – 8 = 9, 9

53 (p. 31 ~ 32)

① 13 ② 15 ③ 2 ④ 9 ⑤ 3
⑥ 10 ⑦ 1 ⑧ 8 ⑨ 1 ⑩ 8
⑪ 15 ⑫ 3 ⑬ 10 ⑭ 13 ⑮ 12
⑯ 2 ⑰ 9 ⑱ 8 ⑲ 16 ⑳ 11

54 (p. 33 ~ 34)

① 14 ② 11 ③ 13 ④ 10 ⑤ 9

⑥ 15	⑦ 8	⑧ 12		
⑨ 17 – 7 = 10, 10	⑩ 17 – 9 = 8, 8			

55	(p. 35 ~ 36)			
① 15	② 12	③ 16	④ 11	⑤ 14
⑥ 17	⑦ 13	⑧ 10	⑨ 9	⑩ 8
⑪ 13	⑫ 16	⑬ 4	⑭ 14	⑮ 5
⑯ 10	⑰ 13	⑱ 1	⑲ 11	⑳ 9

56	(p. 37 ~ 38)			
① 15	② 16	③ 17	④ 13	⑤ 14
⑥ 9	⑦ 12	⑧ 10	⑨ 11	
⑩ 18 – 6 = 12, 12	⑪ 18 – 7 = 11, 11			

57	(p. 39 ~ 40)			
① 7	② 15	③ 3	④ 13	⑤ 4
⑥ 14	⑦ 1	⑧ 9	⑨ 10	⑩ 5
⑪ 15	⑫ 12	⑬ 6	⑭ 16	⑮ 14
⑯ 3	⑰ 13	⑱ 10	⑲ 1	⑳ 11

58	(p. 41 ~ 42)			
① 15	② 10	③ 13	④ 12	⑤ 11
⑥ 14	⑦ 9	⑧ 16		
⑨ 18 – 9 = 9, 9	⑩ 18 – 8 = 10, 10			

59	(p. 43 ~ 44)			
① 12	② 13	③ 1	④ 8	⑤ 9
⑥ 11	⑦ 12	⑧ 14	⑨ 15	⑩ 12
⑪ 1	⑫ 11	⑬ 14	⑭ 13	⑮ 16
⑯ 15	⑰ 10	⑱ 11	⑲ 9	⑳ 8

60	(p. 45 ~ 46)			
① 12	② 12	③ 14	④ 6	⑤ 10
⑥ 14	⑦ 13	⑧ 9	⑨ 15	
⑩ 17 – 8 = 9, 9	⑪ 18 – 7 = 11, 11			

Week 3

61	(p. 49 ~ 50)			
① 18	② 12	③ 16	④ 2	⑤ 7
⑥ 11	⑦ 3	⑧ 13	⑨ 10	⑩ 2
⑪ 12	⑫ 11	⑬ 15	⑭ 6	⑮ 15
⑯ 7	⑰ 17	⑱ 5	⑲ 10	⑳ 14

62	(p. 51 ~ 52)			
① 16	② 10	③ 17	④ 18	⑤ 11
⑥ 14	⑦ 12	⑧ 15	⑨ 13	
⑩ 19 – 5 = 14, 14	⑪ 19 – 8 = 11, 11			

63	(p. 53 ~ 54)			
① 15	② 16	③ 13	④ 11	⑤ 10
⑥ 12	⑦ 17	⑧ 14	⑨ 5	⑩ 15
⑪ 11	⑫ 18	⑬ 10	⑭ 14	⑮ 3
⑯ 8	⑰ 12	⑱ 16	⑲ 8	⑳ 17

64	(p. 55 ~ 56)			
① 17	② 12	③ 10	④ 15	⑤ 14
⑥ 16	⑦ 13	⑧ 11		
⑨ 19 – 7 = 12, 12	⑩ 19 – 6 = 13, 13			

65	(p. 57 ~ 58)			
① 18	② 19	③ 8	④ 18	⑤ 6
⑥ 16	⑦ 5	⑧ 10	⑨ 15	⑩ 7
⑪ 17	⑫ 2	⑬ 12	⑭ 3	⑮ 13
⑯ 4	⑰ 14	⑱ 1	⑲ 11	⑳ 16

66	(p. 59 ~ 60)			
① 14	② 18	③ 19	④ 11	⑤ 16
⑥ 13	⑦ 17	⑧ 12	⑨ 15	
⑩ 20 – 5 = 15, 15	⑪ 20 – 7 = 13, 13			

67	(p. 61 ~ 62)			
① 9	② 19	③ 7	④ 17	⑤ 5
⑥ 15	⑦ 3	⑧ 13	⑨ 2	⑩ 12
⑪ 17	⑫ 1	⑬ 11	⑭ 15	⑮ 19
⑯ 4	⑰ 14	⑱ 18	⑲ 3	⑳ 13

68	(p. 63 ~ 64)			
① 16	② 17	③ 12	④ 13	⑤ 18
⑥ 15	⑦ 11	⑧ 14		
⑨ 20 – 8 = 12, 12		⑩ 20 – 9 = 11, 11		

69	(p. 65 ~ 66)			
① 8	② 17	③ 18	④ 12	⑤ 13
⑥ 14	⑦ 15	⑧ 16	⑨ 17	⑩ 6
⑪ 15	⑫ 16	⑬ 13	⑭ 14	⑮ 9
⑯ 19	⑰ 16	⑱ 12	⑲ 10	⑳ 11

70	(p. 67 ~ 68)			
① 5	② 14	③ 15	④ 4	⑤ 14
⑥ 12	⑦ 10	⑧ 17	⑨ 11	
⑩ 19 – 5 = 14, 14		⑪ 20 – 6 = 14, 14		

Week 4

71	(p. 71 ~ 72)			
① 10	② 12	③ 3	④ 12	⑤ 7
⑥ 3	⑦ 7	⑧ 8	⑨ 9	⑩ 10
⑪ 8	⑫ 3	⑬ 8	⑭ 3	⑮ 12
⑯ 14	⑰ 9	⑱ 6	⑲ 6	⑳ 2

72	(p. 73 ~ 74)			
① 4	② 7	③ 15	④ 10	⑤ 4
⑥ 3	⑦ 4	⑧ 13	⑨ 6	
⑩ 12 – 7 = 5, 5		⑪ 14 – 5 = 9, 9		

73	(p. 75 ~ 76)			
① 2	② 7	③ 12	④ 6	⑤ 8
⑥ 5	⑦ 9	⑧ 15	⑨ 7	⑩ 7
⑪ 2	⑫ 3	⑬ 3	⑭ 7	⑮ 9
⑯ 10	⑰ 12	⑱ 15	⑲ 16	⑳ 11

74	(p. 77 ~ 78)			
① 6	② 15	③ 6	④ 12	⑤ 7
⑥ 4	⑦ 7	⑧ 9	⑨ 3	
⑩ 20 – 8 = 12, 12		⑪ 15 – 7 = 8, 8		

75	(p. 79 ~ 80)			
① 2	② 11	③ 15	④ 13	⑤ 4
⑥ 6	⑦ 4	⑧ 9	⑨ 6	⑩ 3
⑪ 1	⑫ 12	⑬ 18	⑭ 13	⑮ 6
⑯ 10	⑰ 9	⑱ 8	⑲ 10	⑳ 9

76	(p. 81 ~ 82)			
① 0	② 2	③ 4	④ 4	⑤ 6
⑥ 6	⑦ 8	⑧ 8	⑨ 10	
⑩ 18 – 7 = 11,11		⑪ 15 – 7 = 8, 8		

77	(p. 83 ~ 84)			
① 9	② 5	③ 1	④ 6	⑤ 4
⑥ 11	⑦ 3	⑧ 3	⑨ 9	⑩ 3
⑪ 10	⑫ 3	⑬ 8	⑭ 7	⑮ 12
⑯ 7	⑰ 13	⑱ 9	⑲ 6	⑳ 3

78	(p. 85 ~ 86)			
① 2	② 1	③ 5	④ 8	⑤ 12
⑥ 13	⑦ 13	⑧ 6	⑨ 3	
⑩ 16 – 3 = 13, 13		⑪ 17 – 5 = 12, 12		

79	(p. 87 ~ 88)			
① 8	② 16	③ 2	④ 8	⑤ 3
⑥ 8	⑦ 7	⑧ 5	⑨ 11	⑩ 3
⑪ 13	⑫ 1	⑬ 8	⑭ 12	⑮ 7
⑯ 9	⑰ 5	⑱ 13	⑲ 8	⑳ 5

80	(p. 89 ~ 90)			
① 2	② 5	③ 12	④ 9	⑤ 8
⑥ 2	⑦ 9	⑧ 6	⑨ 4	
⑩ 19 – 3 = 16, 16		⑪ 12 – 5 = 7, 7		

Tiger Math

ACHIEVEMENT AWARD

THIS AWARD IS PRESENTED TO

____

(student name)

FOR SUCESSFULLY COMPLETING

TIGER MATH LEVEL C – 2.

Dr. Tiger

____

Dr.Tiger